READING POWER

Earth Rocks!

Sedimentary Rocks

Holly Cefrey

The Rosen Publishing Group's
PowerKids Press™
New York

Published in 2003 by The Rosen Publishing Group, Inc.
29 East 21st Street, New York, NY 10010

First Edition

Book Design: Mindy Liu

Photo Credits: Cover, p. 19 © Corbis; pp. 4–5 © Pat O' Hara/Corbis; p. 4 (inset) © George D. Lepp/Corbis; p. 6 (illustration) Mindy Liu; p. 6 (earth) © Photodisc; p. 7 © Jonathan Blair/Corbis; pp. 8–9 © Howie Garber/Animals Animals; p. 8 (inset) © James P. Blair/National Geographic Image Collection; pp. 10–11 © Kevin Fleming/Corbis; pp. 12–13 © Roger Ressmeyer/Corbis; p. 13 (inset) © Jerome Wyckoff/Animals Animals; pp. 14–15 © Charles O'Rear/Corbis; p. 16 (top) © Vanni Archive/Corbis; p. 16 (bottom) © Larry Lee Photography/Corbis; p. 17 © Jeremy Horner/Corbis; p. 18 © Breck P. Kent/Animals Animals; p. 19 (illustration) Mindy Liu; pp. 20–21 © Tom Bean/Corbis

Library of Congress Cataloging-in-Publication Data

Cefrey, Holly.
Sedimentary rocks / Holly Cefrey.
    v. cm. — (Earth rocks!)
Includes bibliographical references and index.
Contents: Earth rocks — Finding sedimentary rocks — How sedimentary rocks form — Using sedimentary rocks — Rocks that show time.
ISBN 0-8239-6465-5 (library binding)
1. Rocks, Sedimentary—Juvenile literature. [1. Rocks, Sedimentary.]
I. Title.
QE471 .C43 2003

                          2002000161

Contents

Sedimentary Rocks!

There are many kinds of sedimentary rocks. They come in many colors, shapes, and sizes. Some sedimentary rocks are more than 3.5 billion years old! Some are more than 40,000 feet thick.

Shale is made of mud that has been tightly pressed for many years.

Sandstone can form beautiful canyons.

Earth is made of layers of rock.
The top layer is called the crust.
Sedimentary rocks are in, and on
top of, the crust. They are the most
common rocks on the earth's surface.
Sedimentary rocks also make up most
of the ocean floor.

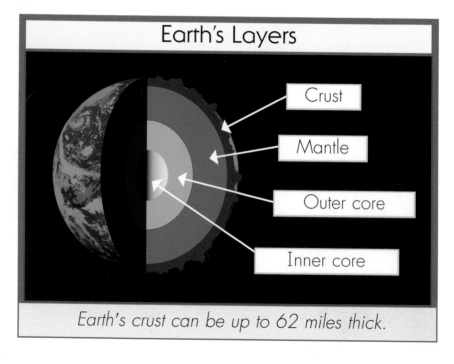

Earth's Layers

Crust

Mantle

Outer core

Inner core

Earth's crust can be up to 62 miles thick.

The Fact Box

Shale is the most common sedimentary rock.

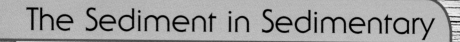

The Sediment in Sedimentary

Most sedimentary rocks are made from sediment. Sediment is bits of sand, rocks, shells, and dirt. Sediment can be found in many places. It is at the bottom of rivers or oceans, in deserts, and in valleys. Sedimentary rocks are also made from animals and plants that have died.

Coal

The Fact Box

Coal is made from plants that have been pressed between rocks for millions of years.

We can see the many layers of desert sand that made this sandstone.

How Sedimentary Rocks Are Formed

It takes many years for sedimentary rocks to form. Year after year, sediment gets covered with more sediment. The top layers of sediment press down on the bottom layers. After millions of years, the tightly-packed sediment becomes rock.

The Fact Box

The Grand Canyon is about one mile deep. The bottom layers are about two billion years old!

Much of the Grand Canyon was made by the Colorado River. As the water moved quickly down the river, it cut away the layers of rock. It carried about 500,000 tons of sediment through the canyon each day.

Mud, clay, and minerals act like glue to hold the pieces of sediment together. If the glue is very strong, the sediment becomes very hard rock. Sediment that is not held together tightly makes a weaker rock, which can break easily.

Some shale is very weak. It can break when you touch it.

The glue in some sedimentary rocks is so strong that it can hold large pieces of rock together.

13

Some sedimentary rocks are formed by minerals. When water in lakes, ponds, or streams dries up, minerals are left behind. These minerals build up and form rocks. Minerals can also build up on shells and other pieces of sediment.

Salt flats are large areas of land where water has dried up, leaving behind salt minerals. These minerals form rock salt, also called table salt.

Using Sedimentary Rocks

There are many uses for sedimentary rocks. Sedimentary rocks are used to make cement, bricks, and tiles. Many famous buildings were made from sedimentary rock.

The Romans used limestone to build the Pantheon.

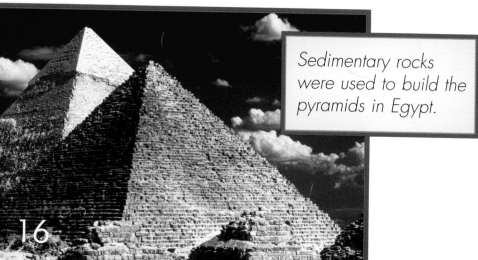

Sedimentary rocks were used to build the pyramids in Egypt.

This woman in Ecuador uses a pick to get limestone. The limestone will be used to make cement, which is used to make houses and buildings.

Studying Sedimentary Rocks

Many geologists study sedimentary rocks. They study fossils, minerals, and other sediments in the rocks. These sediments can show where there were floods or what animals and plants once lived on Earth. Most of the fossils on Earth are found in layers of sedimentary rock.

From studying fossils, scientists have learned that this animal lived more than 66 million years ago.

Layers of Sedimentary Rock in the Grand Canyon

	Number of Years Old
Limestone	265 million
Sandstone	270–275 million
Shale	280 million
Limestone	340–520 million
Shale	540 million
Sandstone	560 million

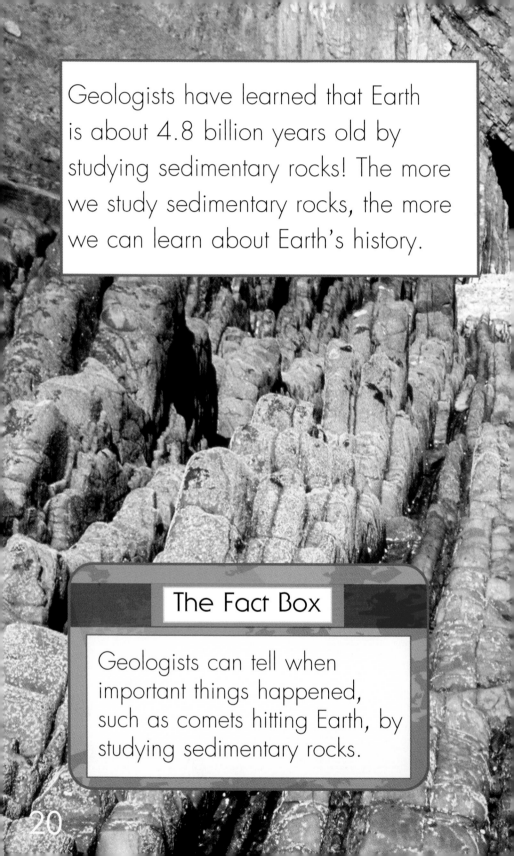

Geologists have learned that Earth is about 4.8 billion years old by studying sedimentary rocks! The more we study sedimentary rocks, the more we can learn about Earth's history.

The Fact Box

Geologists can tell when important things happened, such as comets hitting Earth, by studying sedimentary rocks.

Geologists have learned that the rocks in Devon, England, are 320 million years old.

Glossary

canyon (**kan**-yuhn) a narrow valley with high sides, usually with a stream at the bottom

comets (**kahm**-ihts) very large objects in space that look like stars, with cloudy tails of light

crust (**kruhst**) the solid outer part of a planet

fossil (**fahs**-uhl) the mark or remains of a plant or animal that lived a long time ago

geologists (jee-**ahl**-uh-jihsts) people who study Earth's rocks

layer (**lay**-uhr) one thickness or level of something that is on top of another

minerals (**mihn**-uhr-uhlz) solid matter that comes from the earth

pyramids (**pihr**-uh-mihdz) large buildings in which ancient Egyptians buried their kings

sediment (**sehd**-uh-mehnt) bits of rock, sand, dirt, and shells that settle in bodies of water, deserts, and valleys

sedimentary rock (sehd-uh-**mehn**-tuhr-ee **rahk**) rock that is formed by layers of sediment, which are being pressed together

surface (**ser**-fihs) the outside of anything

Resources

Books

Dig It! How to Collect Rocks and Minerals
by Susan Mondshein Tejada
Reader's Digest Children's Publishing (2001)

Rocks & Minerals
by Jack Challoner
Southwater Publishers (2000)

Web Sites

Due to the changing nature of Internet links, PowerKids Press has developed an online list of Web sites related to the subjects of this book. This site is updated regularly. Please use this link to access the list:

http://www.powerkidslinks.com/ear/sed/

Index

Word Count: 397

Note to Librarians, Teachers, and Parents

If reading is a challenge, Reading Power is a solution! Reading Power is perfect for readers who want high-interest subject matter at an accessible reading level. These fact-filled, photo-illustrated books are designed for readers who want straightforward vocabulary, engaging topics, and a manageable reading experience. With clear picture/text correspondence, leveled Reading Power books put the reader in charge. Now readers have the power to get the information they want and the skills they need in a user-friendly format.